Hyaluronic Acid

Powerful Antiarthritic and Antiwrinkle Supplement

Martin Stone, MH

Copyright © 2007 by Martin Stone

All rights reserved. No part of this publication may be reproduced, stored in a retrieval system, or transmitted in any form without the prior written permission of the copyright owner.

For permissions, ordering information, or bulk quantity discounts, contact: Woodland Publishing, 448 East 800 North, Orem, Utah 84097

Visit our Web site: www.woodlandpublishing.com
Toll-free number: (800) 777-BOOK

The information in this book is for educational purposes only and is not recommended as a means of diagnosing or treating an illness. All matters concerning physical and mental health should be supervised by a health practitioner knowledgeable in treating that particular condition. Neither the publisher nor the author directly or indirectly dispenses medical advice, nor do they prescribe any remedies or assume any responsibility for those who choose to treat themselves.

Cataloging-in-Publication data is available from the Library of Congress.

ISBN-13: 978-1-58054-458-0
ISBN-10: 1-58054-458-4

Printed in the United States of America

07   08   09   10   1   2   3   4   5   6   7   8   9   10

Contents

Introduction 5

Statistics on Aging 5

The Battle Against Time 7

Hyaluronic Acid and Osteoarthritis 15

Research 16

Topical Hyaluronic Acid 21

Hydration and Hyaluronic Acid 21

Conclusion 22

References 23

Introduction

As a society, we are obsessed with youth, youth culture, and slowing the aging process. Most approaches to forestalling the inevitable have been intrusive and cosmetic: plastic surgery, liposuction, Botox injections, and so on. Why are we so concerned with staying young and appearing young? Is it because we've seen the effects of aging on our parents and don't want to end up like them, old and worn out before their time, or perhaps we're afraid of dying in the prime of our life?

Today, more people are living longer than at any other time in history. But still, too many people are dying at too young an age. Do our bodies somehow have a genetically predetermined expiration date no matter what strategies we use to postpone the inevitable signs of age? Are there ways for us to reset this "expiration date" and live longer, healthier lives?

Statistics on Aging

During the second half of the twentieth century, twenty years were added to the average person's life span. Several factors are responsible for this increase in longevity, including advances in medicine, better nutrition, and improved sanitation. These circumstances set the stage for a never-before-seen shift in population dynamics—the baby boom. The previous factors, combined with an increased fertility rate in many Western countries immediately following World War II, has led to a significant increase in the number of people over sixty-five. This phenomenon will continue through 2030. Additionally, the average life span worldwide is expected to increase by another ten years by 2050.

At first glance, this would seem like a wonderful turn of events, but there is a downside. The elderly will require more medical care as they continue to age. Chronic degenerative diseases have a much greater incidence in older populations. Arthritis, heart disease, and cancer contribute to overall disability and a reduced quality of life. As Medicare and other health-related costs increase and take more of the average income, it becomes more imperative for all of us to do whatever we can to reduce or delay the need for long-term care and dependence on public health services.

The following data is from the United States Census Bureau:

- The proportion of the population aged sixty-five and over is projected to increase from 12.4 percent in 2000 to 19.6 percent in 2030.

- The number of people aged sixty-five and over is expected to increase from approximately 35 million in 2000 to an estimated 71 million in 2030.

- The number of people aged eighty years and over is expected to more than double, from 9.3 million in 2000 to 19.5 million in 2030.

- In 1995, the most populous states had the largest number of older people; nine states (California, Florida, Illinois, Michigan, New Jersey, New York, Ohio, Pennsylvania, and Texas) each had more than one million residents who were sixty-five years old.

- In 1995, 15 percent of the population of four states was sixty-five years old; Florida had the largest proportion at 19 percent.

- By 2025, the proportion of Florida's population that's sixty-five years and over is projected to be 26 percent. Fifteen percent in forty-eight states will be sixty-five years old.

The Battle Against Time

Many people know about the effectiveness of glucosamine, chondroitin, omega-3 essential fatty acids, devil's claw, and other supplements used to fight inflammation and treat arthritis. But are there other supplements that can help ameliorate the aches and pains and other effects of aging?

Hyaluronic acid

What if there were a supplement that could help give you:

- Younger looking skin with fewer wrinkles
- Healthier joints with less pain and greater range of motion
- Increased immune system function
- A more youthful appearance

This is the potential of hyaluronic acid, an essential material that the body creates in abundance during our early years but which steadily decreases in quantity as we age.

What is hyaluronic acid?

Hyaluronic acid was first used commercially as a food source. In 1942, Endre Balazs applied for a patent to use it as a substitute for egg whites in bakery products. I'm sure that he never dreamed how many ways hyaluronic acid would be used today.

Glycosaminoglycan can be found in almost all living organisms that have joints and connective tissue. The human body contains mostly water, but we need some way to keep the water in our cells and tissues. This is where hyaluronic acid comes in. While hyaluronic acid has many functions, its primary task is to bind water molecules to cells and tissues, which helps provide the medium our bodies need for molecular transport and numerous other processes.

As our bodies lose the ability to hold water in cells and tissues as we age, joint conditions such as arthritis start to appear. While this is one of the more obvious symptoms of a decrease in hyaluronic acid levels, another symptom of aging appears in our skin—wrinkles. Strangely enough, many people will accept joint pain and reduced mobility as a natural consequence of aging but will do all they can to reduce the appearance of wrinkles. For this reason alone, many people will use hyaluronic acid.

Yuzuri Hara, hyaluronic acid, and long life

In our search for the answer to increased longevity and vitality, ABC aired the television show "The Village of Long Life: Could Hyaluronic Acid Be an Anti-Aging Remedy?" and brought hyaluronic acid to the forefront of public consciousness. The show covers a Japanese village in which a disproportionate number of people live to be ninety years and older. These elderly villagers live longer and have more energy and vitality than the average North American fifty years their junior. A large percentage of these seniors are incredibly healthy, vital, and strong to the extent that they could work every day in the fields and complete laborious tasks. What is the villagers' secret? Diet.

Their traditional diet includes local root vegetables and starches with a high nutritional content that improves the body's natural production of hyaluronic acid. With this increased concentration of hyaluronic acid, the villagers' skin retains its moisture, appearing youthful and wrinkle-free. The villagers exhibit very few signs of vision problems; their eyes are bright and healthy. This is in spite of the fact that they work outside in the sun every day, and in quite a few cases are heavy smokers—all factors known to increase the aging process due to the generation of free radicals.

Despite these facts, over 10 percent of the residents of Yuzuri Hara are over eighty-five. This is approximately ten times the national average found in the United States in any given population. The Yuzuri Hara seniors live longer and are so healthy that they rarely see a doctor and rarely succumb to chronic degenerative diseases found in the United States, such as cancer, Alzheimer's disease, and diabetes.

This widely publicized discovery eventually motivated a large Japanese pharmaceutical firm to develop the first hyaluronic acid pills.

Hyaluronic acid and connective tissue disorders

One of the functions of connective tissue is to lubricate and cushion our joints. It also helps connect the skin to supporting tissues throughout the body. When we are young, our skin is elastic and wrinkles are negligible because hyaluronic acid levels are high enough to maintain adequate water levels in our cells and tissues.

When hyaluronic acid levels are too low, connective tissue disorders typically follow. They include:

- Temporomandibular joint (TMJ)
- Keratoconus
- Mitral valve prolapse
- Osteoarthritis

The properties of hyaluronic acid are wide-ranging and can affect numerous body symptoms. We now know many of the properties of hyaluronic acid and it's specific functions in the body.

- Hyaluronic acid is a special mucopolysaccharide that is created and utilized as a lubricant for human joints. It is found in substantial amounts throughout the body, especially in the young, but is substantially reduced as the aging process occurs.
- When present in a joint, even a joint with minimal or no cartilage, it can provide a cushioning effect, reducing damage and maintaining normal joint function. One of the properties of hyaluronic acid that allows this is its ability to absorb up to three thousand times its own weight in water. Because of its lubricating and joint cushioning properties, further

animal studies have reported potential disease-modifying effects and possible treatments for rheumatoid arthritis and osteoarthritis.

- New research has revealed that hyaluronic acid may stimulate the immune system, activate white blood cells, and control cell migration. This research indicates that hyaluronic acid may reduce the need for antibiotics.

- In addition, hyaluronic acid reduces the growth rate of several strains of bacteria, and it has been reported to reduce the incidence of chronic bronchitis.

- As a result of its primary property of water retention and absorption, hyaluronic acid is now used in cosmetics such as makeup and moisturizing creams to help hold water in the skin and thereby reduce the appearance of wrinkles.

- Hyaluronic acid has also been used during cataract surgery to protect the corneal endothelium during postoperative recovery.

Hyaluronic acid and connective tissue

Connective tissue is what holds us together. Connective tissue suspends and surrounds most of the organs inside the body and is also involved in the delivery of nutrients to tissues and organs. There's so much connective tissue in our bodies that if we removed everything else and just left the connective tissue, we would be able to recognize ourselves in the mirror. Cartilage, bone, and blood are all considered specialized forms of connective tissue, which gives some idea as to the variety of forms that connective tissue can take.

Connective tissue is also found between the cells that make up our body. It gives our tissues substance, form, and strength. Hyaluronic acid is an important substance found in connective tissue and can be compared to the mortar that holds a brick wall together. Just as mortar is composed of several materials, including cement, sand, and water, so too is our connective tissue. Without hyaluronic acid, connective tissue is severely compromised and cannot do its job properly, if at all, which becomes apparent in chronic connective tissue disorders that increase with age.

As you can see, connective tissue is involved in more than just skin formation and wrinkles. Either directly or indirectly, connective tissue is critical to the pain-free and efficient functioning of almost every part of our body, from wound healing to skin formation and joint function.

Connective tissue disorders that involve hyaluronic acid abnormalities include: Ehlers-Danlos syndrome, Marfan syndrome, Osteogenesis imperfecta, and Stickler syndrome.

While these specific conditions primarily result from genetic damage and are often found only in certain families, this is just a small sampling of connective tissue disorders. In every study of connective tissue disorders that examined hyaluronic acid, the levels of hyaluronic acid were always abnormal.

Other connective tissue disorders include:

- Heart valve abnormality, such as mitral valve prolapse
- Joint instability
- Uncontrolled or spastic muscle contraction
- Osteoarthritis
- TMJ
- Acrogeria (premature wrinkling of the skin)
- Fibromyalgia
- Premature aging syndromes
- Glaucoma
- Detached retinas
- Abnormal skeletal formation or hypermobility

Many factors influence hyaluronic acid levels. Genetic factors play a role, but current research is revealing that environmental and nutritional factors also have an impact, including magnesium and zinc levels.

Low levels of magnesium and zinc are also apparent in other conditions such is poor wound healing and mitral valve prolapse. There's a strong probability that these mineral deficiency conditions directly or indirectly affect hyaluronic acid production.

What does the research say about the relationship between zinc and magnesium deficiencies and hyaluronic acid?

- Hyaluronic acid becomes abnormally elevated in the skin of swine with zinc deficiencies. While this may sound like a positive effect, it is manifested as dermatitis.

- Magnesium is needed for hyaluronic acid synthesis. Magnesium supplementation is an established treatment for many of the symptoms of connective tissue disorders, such as fibromyalgia, mitral valve prolapse, and contractures.

- Vitamin C can degrade hyaluronic acid. Large levels of dietary vitamin C can reduce the molecular weight of hyaluronic acid by two-thirds, reducing its effectiveness as a water-retaining molecule and its ability to act as a joint lubricant.

- Estrogen replacement therapy increases hyaluronic acid activity. In some experimental studies, estrogen increased the activity of fibroblasts and water and hyaluronic acid and collagen skin contents, which is why estrogen is prescribed for thinning of vaginal and other delicate tissues in menopausal women. Estrogen is known to increase the effective use of nutrients like magnesium and zinc, the same nutrients that are known to affect hyaluronic acid levels.

- Cigarette smoke is known to degrade hyaluronic acid. Smoke reduces the chain length of purified hyaluronic acid and reduces the viscosity of the proteoglycans exposed to the altered hyaluronic acid, which strongly alters its lubricating and cushioning properties.

- In a study of rats, hyaluronic acid turnover and metabolism were affected by age, dietary composition, and caloric intake. Calorie restriction maintains hyaluronic acid levels similar to those of young rats throughout life in this animal study. Hyaluronic acid levels in twelve- to twenty-month-old rats were only 10 to 29 percent of the levels in both younger and elderly rats when they were allowed to eat as much as they pleased. The more they ate, the less hyaluronic acid was produced in the joints. Hyaluronic acid levels were also measured in rats fed a semisynthetic diet in which the protein source was hydrolyzed casein. Hyaluronic acid levels differed from four to twenty fold at every age examined. Caloric restriction also positively affected hyaluronic acid levels in nineteen-month-old casein-fed rats; hyaluronic acid levels were 2.3 times higher than rats the same age fed a normal diet and were not substantially different from young or aged animals.

    Since this animal study showed good results in calorie restriction, this could be a clue that diet may affect hyaluronic acid levels in humans as well. In another study on rats, hyaluronic acid deposition in rat cerebellums (brain tissue) is affected by thyroid deficiency, thyroxine treatment, and undernutrition.

- The change in hyaluronic acid levels is apparent in the smaller number of brain cells formed in undernourished rats, or alterations of cell maturation (accelerated in thyroxine-treated and slowed down in thyroid-deficient rats). This animal study reported reduced brain cell development and loss of tissue water, which was associated with a lack of hyaluronic acid in the surrounding tissue as a result of abnormally low levels of thyroxine and poor nutrition.

- Animal studies have reported that a lack of adequate nutrition over a long period of time also strongly affects collagen formation, just as collagen formation and concentration dysfunctions are found in many connective tissue disorders. Whether this is a direct result of a lack of magnesium or zinc due to nutritional deficiency is unclear. While we know that a deficiency of these two minerals directly affects hyaluronic acid production, we don't know if there are other nutritional substances that also contribute to overall hyaluronic acid deficiency.

- Strep and staph bacteria produce an enzyme called hyaluronidase, which breaks down hyaluronic acid, thus allowing a point for the bacteria to enter the body. This may be why people may become hypermobile (loose joints) or develop heart ailments like mitral valve prolapse after illnesses such as rheumatic fever, because the hyaluronic acid in their connective tissue has been degraded by the bacteria that caused their illness.

- If animals that are genetically similar to humans such as rats can have reduced collagen levels and hyaluronic acid abnormalities from changes in their diets, then it would be reasonable to consider diet as a causative factor in human hyaluronic acid abnormalities.

One point to consider is that the mineral content in our soils has been low to nonexistent for almost seventy years. This is a direct result of industrial levels of food production that have depleted mineral content in all arable soils found in the United States. If we can't get mineral content in the foods we eat due to mineral-depleted soil, it's no wonder that zinc and magnesium deficiencies abound in the Western world. This could be one of the major contributing factors to a further reduction of hyaluronic acid production as we age.

Hyaluronic Acid and Osteoarthritis

It has been said that osteoarthritis will affect almost everyone if they live long enough. This degenerative joint disease is the most common form of arthritis. Once thought to be a condition found mostly in the elderly, osteoarthritis is being found more commonly in people as young as teenagers.

The most common symptom of osteoarthritis is cartilage breakdown. Currently, X-ray studies are the most common tool used to diagnose osteoarthritis. Any kind of radiation, whether incidental or therapeutic, carries with it its own risk of free-radical and radiation damage. This is not the only weakness of X-ray diagnosis. Changes in a joint may not be visible by X-ray observation for up to three years after they have begun. This is a large window of opportunity for preemptive treatment before permanent damage occurs.

Tests that recognize the presence of biomarkers can help physicians identify early signs of disease, more reliably detect disease progression, and assess patient response to treatment. Serum hyaluronic acid levels are being studied as a potential sign of the presence and severity of osteoarthritis.

Johnson County Osteoarthritis Study

The Johnson County Osteoarthritis Study included 753 participants and examined several factors in relation to serum hyaluronic acid: X-ray evidence of osteoarthritis, age, gender, race, and body mass index (BMI).

Various self-reported coexisting disease conditions were used as a measurement of hyaluronic acid levels. They included: circulation problems, cancer, gout, high blood pressure, diabetes, and rheumatoid arthritis

Joanne Jordan, MPH, MD, and her colleagues at the University of North Carolina, Chapel Hill, and Duke University, reported that there was a strong correlation between serum hyaluronic acid levels and increasing osteoarthritis severity as measured by X-rays of the knees and hips. Additionally, regardless of disease severity, serum hyaluronic acid levels were generally higher in men compared to women, and in Caucasians compared to African Americans.

Only gout was found to have an independent association with serum hyaluronic acid, possibly as a result of the severe inflammation and joint damage caused by gout.

One surprising aspect of this study was that gender and ethnicity played a role in hyaluronic acid production. More research is needed to explore what underlies the connection between gender and ethnicity and hyaluronic acid levels.

Research

Much of the research on hyaluronic acid has been done on injectable forms, in which the hyaluronic acid is injected directly into the affected joints to increase mobility and reduce pain. Most, but not all, studies have reported good results using this method. Some of the varied results of this research are outlined below.

Osteoarthritis of the knee affects up to 10 percent of the elderly population. The condition is frequently treated by intra-articular injection of hyaluronic acid. A systematic review and meta-analysis of randomized controlled trials has been initiated to assess the effectiveness of this treatment.

Twenty-two different studies were identified and included in a systematic review of the efficacy of hyaluronic acid injection on osteoarthritic knees. Measurement of symptoms such as the level of pain at rest, during or immediately after movement, joint function, and adverse side effects were all included. The end result was that pain at rest was improved by hyaluronic acid injection.

Despite the improvement, this report suggests that hyaluronic acid has not been proven to be clinically effective and suggests that larger trials are necessary to fully identify the benefits and risks of hyaluronic acid injections.

Obviously, this meta-analysis of twenty-two different studies did not endorse the use of hyaluronic acid as a treatment for osteoarthritis. However, the story does not end here.

Other studies reported very different results using hyaluronic acid injections in treating osteoarthritis.

In one study, participants who had never used hyaluronic acid injections were investigated. These patients received a series of three injections over a period of three weeks. The results were very different frrom the previous studies. Researchers reported that hyaluronic acid injections were very effective in reducing overall arthritic pain and were highly effective in relieving resting and walking pain after two treatment series. The patients were very satisfied with the therapy and experienced very few local adverse events. Reduced use of other pain-reduction modalities, including drug therapy, was also reported. These data support the potential role of intra-articular hyaluronic acid as an effective long-term therapeutic option for patients with osteoarthritis of the knee.

While this study showed improvement over a six-month period, what about short-term treatment effectiveness? Another study was instituted to evaluate the efficacy and tolerability of a course of five injections of hyaluronic acid given at intervals of one week in patients with symptomatic mild to moderate osteoarthritis of the knee.

This eighteen-week double-blind study included five injections administered at one week intervals. The effects of the injections were monitored for a further thirteen weeks.

Two hundred forty participants received injections of either 25 milligrams of hyaluronic acid or a placebo. Of the 240 initial participants, 223 were included in the final analysis after all the data were collected. Once again, the hyaluronic acid–treated group had a statistically different outcome compared to the placebo group.

There were significant improvements in pain and stiffness overall, but these did not become apparent until after the series of injections was completed. This study reported no serious adverse events associated with hyaluronic acid injections.

According to the next study, hyaluronic acid injections work as well as knee exercise to increase joint mobility. The goal of this study was to determine whether hyaluronic acid or progressive knee exercises can improve mobility in patients with osteoarthritis of the knee.

In this study, 105 participants, all with approximately the same degree of osteoarthritis of the knee, received either three intra-articular hyaluronic acid injections over three weeks or were given physical exercise and rehabilitation over six weeks. The effects of these therapies were evaluated after eighteen months.

Surprisingly, there was very little difference between the two groups, with substantial improvement in both groups even after eighteen months. This study only measured joint mobility and didn't include pain reduction.

Although physical exercise may have increased the mobility of the joint, it was still a dry joint and would be painful to use, whereas the hyaluronic acid test subjects increased their joint mobility as a result of reduced pain in movement, allowing them to increase joint movement naturally.

Another study reported that hyaluronic acid injections resulted in better patient outcomes in people with milder forms of osteoarthritis and in those who started treatment early in the progression of the disease. The later you start hyaluronic acid injections and the more advanced the joint damage is, the less effective hyaluronic acid will be.

The primary objective of the next study was to investigate structural changes, as measured by joint space narrowing, within the knee joint during treatment with intra-articular injection of hyaluronic acid of molecular weight 500–730 kilodaltons in patients with osteoarthritis of the knee.

As osteoarthritis advances, the joint space becomes narrower until bone rests on bone with no cartilage left to cushion them.

This double-blind placebo-controlled human study included the use of regular weekly injections of either hyaluronic acid or placebo over three weeks. Treatment with pain medications was allowed. The only measurement of success or failure was the reduction in the joint space width or joint space narrowing. A total of 408 patients were randomized, and 319 completed the one-year study.

In patients with radiologically more severe disease, there was no difference in joint space narrowing between the two treatments.

Although, in this one-year study, no overall treatment effect was seen, those with radiologically milder disease at the start of the study had less progression of joint space narrowing when treated with hyaluronic acid.

In another study, 210 patients aged sixty years and older with osteoarthritis of the knee were treated with three injections weekly over a three-week period. Two different hyaluronic acid products were used along with a saline placebo with the primary measures of effectiveness used being weight-bearing pain. The intra-articular injections produced a significant reduction in weight-bearing pain, resting pain, and maximum pain over twenty-six weeks.

There were no significant differences in outcome between any of the three study groups during the first twenty-six weeks. Even the placebo saline injection worked to some degree.

Patients with osteoarthritis of the knee who were treated by injection into the knee of either of two hyaluronic acid preparations or placebo showed clinical improvements during the first twenty-six weeks of treatment, though neither preparation offered a longer duration of clinical benefit than placebo. However, when data for the two treatments were pooled, there was a significantly longer duration of clinical benefit for hyaluronic acid treatment than for placebo, indicating that there is a difference in the treatment effectiveness between the two hyaluronic acid products tested.

Research is ongoing. The goal of developing effective topical and oral products is coming closer. These products would enable patients to effectively treat themselves at home without the need for a visit to the doctor's office for an injection.

A recent clinical study published in *The Federation of American Societies for Experimental Biology Journal* confirms that sternal hydrolyzed collagen type II combined with hyaluronic acid is safe and effective in relieving pain and stiffness from osteoarthritis and encourages overall joint health in adults.

The holy grail of hyaluronic acid research for many years has been to find a topical or oral application that is easily absorbed. For some time, studies have reported that hyaluronic acid is not easily absorbed through the skin or the gastrointestinal tract. But new research has reported that an oral form with significant absorption and bioavailability in normal volunteer subjects has been developed.

Until now, regular or native collagen (which has not been predigested but does contain hyaluronic acid) that was comprised of molecules that were much too large for easy absorption in the gastrointestinal tract was the only form available.

The key to increasing absorption was to reduce the typical molecular weight of collagen and hyaluronic acid. Reducing the molecular weight to between 1,500 and 2,500 daltons substantially increased absorption and bioavailability.

By determining the rate and magnitude of hyaluronic acid absorption and its bioavailability, this study clearly demonstrates that this particular form of hyaluronic acid has the characteristics needed to move rapidly from blood to tissue.

New research conducted by the Federation of American Societies for Experimental Biology, including a consortium of scientists, released clinical research results about oral delivery of radio-labeled hyaluronan and its ability to be taken up by joints. Dr. Alex Schauss, director of the American Institute for Biosocial and Medical Research, presented the findings at the 2004 Experimental Biology conference. The study demonstrated that hyaluronic acid can be absorbed effectively through the intestine and into the blood in substantial amounts.

This research is some of the earliest suggesting that hyaluronic acid can be absorbed orally, paving the way for hyaluronic acid dietary supplements to be introduced for arthritis treatment and as anti-wrinkle agents.

Until now, there were no data on absorption levels after oral intake, and the therapeutic use of hyaluronic acid was limited to injection or topical application. The results of this study, which examined the absorption, excretion, and distribution of radio-labeled hyaluronic acid after a single oral administration in genetically modified Wistar rats and beagle dogs, demonstrated that hyaluronic acid is absorbed and distributed to organs and joints after a single oral administration.

Topical Hyaluronic Acid

Recently, a combination of 3 percent diclofenac and 2.5 percent hyaluronic acid has been approved for use in the United States, Europe, and Canada for treatment of actinic keratosis, which is the third most common skin disorder in the United States.

Actinic keratosis can be the first step in the development of skin cancer. It is estimated that up to 10 percent of active lesions, which are redder and more tender than the rest, will take the next step and progress to squamous cell carcinoma (skin cancer). These cancers are usually not life-threatening, provided they are detected and treated in the early stages. However, if the condition is not detected early enough, the lesions can bleed, ulcerate, become infected, and invade surrounding tissues. Three percent of the time, cancerous cells will metastasize and spread to the internal organs.

Hydration and Hyaluronic Acid

An important point to remember is that hyaluronic acid is hydrophilic (water loving). It can capture up to three thousand times its own weight in water to act as a lubricant and cushion for joints and collagen throughout the body if the water is available. Hyaluronic acid can't live up to this potential if you are dehydrated. It is crucial to drink enough filtered water; the formula to use is half your body weight in ounces. If you weigh 150 pounds, drink a minimum of 75 ounces per day on an empty stomach, between meals. Drink more if you live in a dry climate, if you exercise regularly or if you drink more than one cup of coffee or can of soda pop per day.

Conclusion

Hyaluronic acid is proving to be a valuable ally in the treatment of many conditions ranging from arthritis to skin disease. The fact that it has few, if any, side effects and has such long-lasting therapeutic effects makes it a natural addition to any baby boomer's medicine cabinet. Typically, when we can use substances that naturally occur in the body (orthomolecular substances), we don't usually have to worry about side effects (although there are some exceptions, especially hormones).

People are becoming more aware that we don't have to be satisfied with merely covering the symptoms of illness; we can get to the root cause of the condition in many cases by simply giving the body the nutrients it needs to rebuild itself; this is what every system in the body does every day. This is how we were designed; we heal ourselves from daily damage every day. Think of it this way: when you cut yourself, does the bandage you put on the cut heal the injury? Of course not, it merely reduces the chance of infection. Your body does all the repair work by itself, with no conscious input from you. For this reason, it is important to realize that unless the foundations of health are addressed first, it doesn't matter what supplements you use. You can't replace clean water, a healthy diet, frequent physical exercise, and healthy emotional attitudes with a supplement. The word "supplement" says it all: it is meant to supplement a healthy lifestyle.

References

"A double blind, randomized, multicenter, parallel group study of the effectiveness and tolerance of intra-articular hyaluronan in osteoarthritis of the knee." *J Rheumatol.* 2004 Apr; 31(4): 775–82.

"A one-year, randomised, placebo (saline) controlled clinical trial of 500-730 kDa sodium hyaluronate (Hyalgan, Hyaluronic Acid) on the radiological change in osteoarthritis of the knee." *Int J Clin Pract.* 2003 Jul–Aug; 57 (6): 467–74.

Alexander G., et al. "Absorption, distribution and excretion of 99 m labeled hyaluronan after single oral doses in rats and beagle dogs." Life Sciences Division, American Institute for Biosocial and Medical Research, Inc.

Campbell P.R. "Population projections for states by age, sex, race, and Hispanic origin: 1995 to 2025." U.S. Bureau of the Census, Population Division, PPL-47, 1996.

"Hyaluronic acid injection of the knees is no more effective than placebo." *Archives of Internal Medicine.* 2002; 162: 245–47.

"Hyaluronic Acid: a unique topical vehicle for the localized delivery of drugs to the skin." *J Eur Acad Dermatol Venereol.* 2005 May; 19(3): 308–18.

"Intra-articular hyaluronic acid compared with progressive knee exercises in osteoarthritis of the knee: a prospective randomized trial with long-term follow-up." *Rheumatol Int.* 2005 Mar 18.

"Intra-articular hyaluronic acid for the treatment of osteoarthritis of the knee: systematic review and meta-analysis." *CMAJ.* 2005 Apr 12; 172 (8): 1039–43.

Jordan J.M., et al. "Serum hyaluronan levels and radiographic knee and hip osteoarthritis in African Americans and Caucasians in the Johnston County Osteoarthritis Project." *Arthritis and Rheumatism.* 2005; 52(1): 105–11.

Kinsella K., V. Velkoff. "An Aging World: 2001." U.S. Census Bureau. Washington, D.C.: U.S. Government Printing Office, 2001; series P95/01–1.

McDevitt C.A., Beck G.J., et al. "Cigarette smoke degrades hyaluronic acid." *Lung.* 1989; 167(4): 237–45.

Motohashi N., et al. "The effect of synovial fluid proteins in the degradation of hyaluronic acid induced by ascorbic acid." *J Inorg Biochem.* 1985 May; 24(1): 69–74.

References

Normand G., Vitiello F., et al. "Developing rat cerebellum-II. Effects of abnormal thyroid states and undernutrition on Hyaluronic Acid." *Int J Dev Neurosci.* 1989; 7(4): 329–34.

Petrella R.J. "Hyaluronic Acid for the treatment of knee osteoarthritis: long-term outcomes from a naturalistic primary care experience." *Am J Phys Med Rehabil.* 2005; Apr; 84 (4): 278–83.

Prasad A.S., Rabbani P., et al. Experimental zinc deficiency in humans. *Annals of Internal Medicine.* 1978 Oct; 89(4): 483–90.

Thomson R.W., et al. "Alteration of porcine skin acid mucopolysaccharides in zinc deficiency." *J Nutr.* 1975 Feb; 105(2): 154–60.

U.S. Census Bureau. International database. Table 094. Midyear population, by age and sex. U.S. Census Bureau. State and national population projections.

United Nations. Report of the Second World Assembly on Aging. Madrid, Spain: United Nations, April 8-12, 2002.

Vaillant L., Callens, A. "Hormone replacement treatment and skin aging." *Therapie.* 1996 Jan–Feb; 51(1): 67–70.

Yannariello J., Chapman S.H., et al. "Circulating hyaluronan levels in the rodent: effects of age and diet." *Am J Physiol.* 1995 Apr; 268 (4 Pt 1): C952–7.